WHAT WOULD HAPPEN IF...

ALL THE RAINFORESTS WERE DESTROYED?

Written by Izzi Howell

Illustrated by Paula Bossio

WORLD BOOK

www.worldbook.com

READING TIPS

This book asks readers to ponder the question *what would happen if all the rainforests were destroyed?* Readers will discover how rainforests support life across the globe and why they are in danger. They will consider the major problems destroying rainforests will cause and ways we can try and save them. Use these tips to help readers consider the ripple effects of certain actions and events.

Before Reading

Explain to readers that this book uses cause and effect to show how a change in one part of the world can affect the environments and living things throughout the rest of the world. Cause and effect can help us think about why things happen the way they do. It can also help us think about what might happen in the future because of our actions today. Encourage readers to be on the lookout for examples of a cause and effect structure as they explore what would happen if all the rainforests were destroyed.

During Reading

Discuss with readers how some actions and events have multiple causes and others have multiple effects. Explain that it can be tricky to keep all the if/then scenarios straight in our minds, so it can be helpful to create a visual guide. Encourage readers to draw and add notes to their own cause and effect maps like those found on pages 20-21, 26-27, and 31.

After Reading

After finishing the book, discuss with readers how their understandings and opinions of rainforests, biodiversity, and deforestation have changed. Additionally, you can have readers respond to the comprehension questions included on page 46 and can complete the Chain of Events activity on page 47 to further extend the learning.

Visit **www.worldbook.com/resources** for additional, free educational materials.

There is a glossary of terms on pages 44–45. Terms defined in the glossary are in boldface type that **looks like this** on their first appearance on any spread (two facing pages).

Contents

Rainforests at risk 4
Remarkable rainforests 6
Rainforests in danger 12
A new land 18
Going, going, gone 22
Moving out 28
Other problems 30
Save our rainforests! 34
Conclusion 40
Summary 42
Glossary 44
Review and reflect 46

Rainforests at risk

What comes to mind when you think of rainforests? Trees? Wild animals? Yes, of course, but that's not all! Rainforests are some of the most incredible and important **ecosystems** on Earth. They are home to lifesaving medicines and delicious foods, and they even store massive amounts of carbon. Go rainforests!

Many different species of animals, including these blue-and-yellow macaws, live in rainforests.

DID YOU KNOW?

- The Amazon rainforest covers an area twice the size of India!

- The wettest rainforests can receive over 30 feet (9 meters) of rain every year.

- Rainforests are the oldest living ecosystem on Earth. Some have been in the same place for at least 70 million years.

- The small rain-forested country of Panama has more bird species than all the United States and Canada combined.

- The greenery in the rainforest is so thick that it can take 10 minutes for a drop of water to fall from the canopy to the forest floor.

However, despite their huge value, rainforests are in serious danger. Humans have been destroying the rainforests at an ever-increasing rate over the past 500 years—and more so than ever during the past century.

The loss of the rainforests would have a terrible impact on the plants, animals, and people that live there. It would also cause serious problems for the rest of our planet. Let's find out why!

THINK ABOUT IT!

What do you think it would be like to explore a rainforest? Think about what you could see, hear, and smell.

Remarkable rainforests

First things first—what is a rainforest exactly? Well, as the name suggests, a rainforest is a woodland area that receives lots of rain! Plants grow very well in these conditions, and rainforest trees often reach massive heights.

There are two main types of rainforests—tropical and temperate. Tropical rainforests grow near the **equator** in South America, Asia, Africa, and Australia. Because these forests are so near the equator, temperatures there are very warm, and conditions are moist and **humid.**

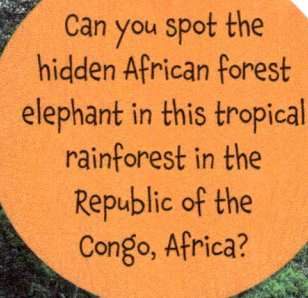

Can you spot the hidden African forest elephant in this tropical rainforest in the Republic of the Congo, Africa?

Temperate rainforests are found outside of the tropics on most continents. They almost always grow along the coast. The weather in temperate rainforests is cool and damp. In this book, we'll mainly be talking about tropical rainforests, so let's forget about temperate rainforests for now!

Did you know that that many of the foods in your kitchen grow in tropical rainforests? Such fruits as bananas, pineapples, and avocados; such spices as black pepper, cinnamon, and vanilla; and such everyday favorites as chocolate and coffee all come from rainforest plants! Many rainforest plants can also be used to make medicine.

Chocolate is made from the seeds of the cacao tree, which grow inside these pods.

DID YOU KNOW?

About one-quarter of all medicines today contain ingredients that come from the rainforest!

THINK ABOUT IT!

If all the rainforests were destroyed, we wouldn't be able to enjoy any more foods that grow there. Which food do you think you would miss the most?

REMARKABLE RAINFORESTS

Tropical rainforests are home to an unbelievable number of different species of animals and plants. In fact, over half of the species found on Earth live in the rainforests! And that's not all. Scientists believe that there are even more rainforest species out there that humans are yet to discover.

One of the reasons rainforests are so **biodiverse** is because they contain many different microhabitats, or layers, at different heights that support different types of plants and animals. The temperature, amount of light, and even the weather change as you move through the layers.

FUN FACT! The Amazon rainforest is home to 2.5 million species of insects, 40,000 different species of plants, and nearly 1,300 species of birds.

Let's head up to the emergent layer!

THINK ABOUT IT! Which layer of the rainforest would you most like to visit? Why?

EMERGENT LAYER

The tops of high trees that tower over the rest of the forest are known as the emergent layer. Up here, the weather is very sunny, but also rainy and windy! Birds, bats, and butterflies make their home up here in the treetops.

CANOPY LAYER

Rainforest trees grow so closely together that their top leafy branches form a type of roof known as the canopy. There are plenty of fruits and nuts to eat up here, so it's a popular place for many animals to live, including sloths, monkeys, and birds.

UNDERSTORY LAYER

Some rain and sunlight make it through to the understory layer, which is made up of shrubs and the lower branches of trees. Many large **predators** hide out in the branches, waiting to pounce!

FOREST FLOOR

The forest floor is the lowest rainforest layer. Very little light makes it through the thick canopy of leaves up above, so it's very dark down here. Lots of insects live among the dead and **decomposing** leaves that have fallen to the floor.

REMARKABLE RAINFORESTS

Rainforest trees also perform a hugely important role that's totally invisible to the naked eye! They store large amounts of the element carbon, which helps limit the levels of **carbon dioxide** gas in the atmosphere. This, in turn, helps reduce **global warming** and **climate change.** Let's take a closer look.

PLANTS AND CARBON

Carbon dioxide in the atmosphere

Excess carbon dioxide can trap heat near Earth's surface, contributing to global warming—an increase in Earth's average surface temperature.

THINK ABOUT IT

Use the information on this page to predict how the destruction of the rainforests is affecting the amount of carbon in our atmosphere. Then check your answer on pages 30-31.

Trees and plants absorb carbon dioxide during **photosynthesis** (the process by which plants make their own food from sunlight). Some of this carbon is stored inside the plant. It will remain there until the plant dies.

When a plant dies, it releases carbon dioxide as it **decomposes.**

Many plants are cut down to be burned as fuel. Burning plants releases the carbon dioxide stored within them.

Rainforests in danger

For most of human history, rainforests were mostly untouched. Many people made their homes in and around rainforests, but they learned how to live in and take from the forests without doing too much harm.

This all changed in the 1500's, when European settlers started exploring and clearing rainforest land in South America. They removed valuable resources from the forests and used the land for farming and development. Since then, rainforest **deforestation** has continued at an alarming rate.

So, why are rain forests destroyed?

DID YOU KNOW? Less than 50 percent of the area covered by tropical rainforests in 1500 still has trees today.

12

It's pretty well-known that rainforests are an incredible and important **ecosystem,** so why are people still destroying them? Well, the answer is usually money. When cleared, rainforest land can be used as farmland, which generates jobs and income. People pay a fortune for valuable wood from rainforest trees, oil, and metal and precious stones from mines. **Hydroelectric dams** generate electricity, which can be sold on for lots of money.

Many rainforests are found in less economically developed areas where some people struggle to get by. The rainforest and the land it stands on provides opportunities for them to make enough money to survive. Other people, however, have plenty of money and still continue to destroy the rainforests. They don't care about the plants, animals, or biodiversity. They just care about money.

Small farms built on rainforest land (left) do less damage than massive farms, such as this palm oil plantation in Southeast Asia (right).

FUN FACT!
Around 1.2 billion people around the world depend on rainforests to make enough money to survive.

RAINFORESTS IN DANGER

THINK ABOUT IT

Which type of person do you think is more of a risk to the rainforests: someone who doesn't have much money and uses rainforest resources to survive, or someone rich who destroys the rainforests to make even more money? Why?

Even people who live far away from rainforests can damage them without realizing it. The entire planet is contributing to **climate change** by burning **fossil fuels,** which releases **greenhouse gases.** These gases gather in our atmosphere and trap heat from the sun. This makes temperatures rise on Earth, which can upset the delicate balance in many ecosystems, including rainforests.

Here's a scary fact—since 1988, about 10,000 acres (4,047 hectares) of just the Amazon rainforest alone have been destroyed every day. That's about 12 times the size of Central Park in New York, U.S., every single day. If we continue to destroy the rainforests at a similar rate, how long will it be before they are gone for good?

Experts believe that many rainforests could disappear altogether within the next hundred years. But long before the end, we'll see many rainforest species become **extinct** due to lack of food and places to shelter. If we continue to destroy rainforests at the same rate, it's estimated that rainforests will lose about 5 to 10 percent of their species every 10 years.

Orangutans live in the rainforests of Borneo and Sumatra in Southeast Asia. They are seriously **endangered** due to **habitat** loss and hunters who capture them to sell as pets.

RAINFORESTS IN DANGER

However, we may see the end of the rainforests sooner than expected. Some scientists believe that we might not need to actually destroy all of the rainforests for them to disappear. If enough rainforest is cleared and damaged, the rest of the **ecosystem** may not be able to recover, and so it will collapse by itself. The trees will die off, and the land will eventually turn into **grassland**.

DID YOU KNOW?

At the moment, about 75 percent of the Amazon is now considered unstable due to **drought** and **wildfires**.

Where's all my fruit gone?

The grassland that replaces the rainforests won't be a good habitat for rainforest animals, but it will be alright for cattle to munch on

THINK ABOUT IT

What do you think would happen if all the rainforests turned to grassland? Make a prediction, and then check your answer as you read the rest of the book.

So what would happen if the rainforests disappeared forever? Read on to find out!

17

A new land

Rainforests are often cleared because people want the land that they stand on. Let's take a look at what would happen to this land after the rainforests were gone.

If left alone, the land that the rainforests previously stood on would slowly turn into **grassland.** Over the following months and years, wild grasses and shrubs would return, along with a few trees. However, it is very unlikely that the rainforest would naturally grow back.

Large areas of these new grasslands would be used as pasture for huge cattle ranches. There's a huge demand for beef around the world, and it's currently the biggest reason for rainforest **deforestation** in South America. So, with the rainforests gone, people would take advantage to set up lots of new farms.

Beef is a popular meat in many countries around the world.

Some land would be used to grow **crops**. The most popular crops to grow on cleared rainforest land include palm oil, soybeans, corn, and cotton. These crops are used to make our food and are also turned into food for animals on farms.

Workers on a palm oil plantation in Indonesia harvest oil palm fruit.

DID YOU KNOW?

Palm oil is used in almost 50 percent of packaged supermarket products, including chocolate, bread, shampoo, toothpaste, cereal, and cookies.

THINK ABOUT IT

Do you think it's right to eat food that is grown or raised on rainforest land? Why or why not?

But the story doesn't end there. What would be the consequences of using cleared rainforest land in these ways?

GROWING CROPS

Rainforest soil is very low in **nutrients**, which makes it pretty bad for farming. Without lots of extra fertilizer, any **crops** planted would quickly use up the nutrients in the soil.

After losing all its nutrients, the land wouldn't be fertile enough to use as farmland. It might eventually become **grassland**, or in the worst-case scenario, a desert where nothing can grow.

CATTLE RANCHES

It may sound strange, but cattle are one of the major contributors to **global warming**! When cattle fart and burp, they release huge amounts of methane, which is a **greenhouse gas.** Because of the massive amount of cattle around the world, this methane adds up!

GRASSLANDS

The new grasslands that replace the rainforests would become the **habitat** of certain wild animals, but not necessarily those that lived in the

A NEW LAND

Dramatically increasing the number of cattle ranches would result in even more methane being released into the atmosphere. This would make the **greenhouse effect** even worse and contribute further to global warming.

DID YOU KNOW?
Cows and other farm animals are responsible for about 14 percent of all climate emissions from human activity.

Sorry!

rainforests before. The grasslands wouldn't have the right kind of food or shelter to support all the animals that previously lived in the rainforest.

The sad truth is that this land wouldn't remain wild for long. It would probably be cleared again to use as farmland or as a space to build new roads or towns.

I can't believe I have to move AGAIN!

Going, going, gone

If the rainforests were destroyed, millions of different species of plants and animals would lose their homes. These species are adapted to life in the rainforest and depend on the forest for food and shelter. So, without the rainforests, they'd be in big trouble!

Many plants and animals would be killed during rainforest clearance. Trees and plants would be chopped down. Huge numbers of plants and animals wouldn't survive in the fires set to clear the land.

I'm out of here!

A few animals would manage to run away from the flames and escape into nearby **habitats**. Others might try to make a new home on the **grasslands** that replaced the rainforests. Most of these animals would struggle to survive in these new areas. **Herbivores** wouldn't be able to find the rainforest plants that they are used to, and carnivores wouldn't be able to find their normal **prey** in a new habitat. Many would die of starvation.

This white-headed capuchin wouldn't be able to find its favorite food (fruit!) in grassland areas.

Some animals would be split up from other animals from their species. They wouldn't be able to find a mate, or **reproduce,** so no new animals would be born to replace those lost during rainforest clearance, or due to old age or illness.

Before long, most rainforest species would become extremely **endangered.** Without the right food or opportunities to mate, their **populations** would continue to fall, and soon they'd become **extinct.** Considering that over half of the species on Earth live in the rainforests, this would be a monumental loss to our planet.

THINK ABOUT IT

Can you think of another animal that is already extinct? Why did it become extinct?

Not the woolly mammoth, think of another one!

23

GOING, GOING, GONE

A few rainforest species might be able to adapt to life in nearby **habitats.** However, their arrival would upset the delicate balance of the **ecosystem.** They'd have to share resources with the animals that already live there, which would probably cause problems. Uh-oh!

Rainforest **predators,** such as jaguars or pythons, would have to compete with existing predators for **prey.** If the new arrivals were better hunters, the predators that already lived there might go hungry and struggle to survive.

If lots of rainforest **herbivores** moved into a new habitat, they'd munch through way more plants than the previous herbivores that lived in the area. This would put a lot of pressure on the food supply for all the herbivores. Some species might experience **population** loss since they wouldn't be able to find enough food.

Tigers live in rainforests across India and Southeast Asia. Habitat loss is pushing them closer to villages and towns, which has resulted in an increase in attacks on people.

As well as the tragic loss of many known rainforest species, scientists believe that there are many species of plants and animals in the rainforest that are yet to be identified. If the rainforests were destroyed, these species would become **extinct** before we ever saw or studied them. Some of these plants could be used for medicine or even as a new treatment for serious diseases, such as cancer.

FUN FACT!

On average, a new species of plant or animal is discovered every three days in the Amazon rainforest.

This tiny frog, known as a pumpkin toadlet, was only discovered in 2021. It lives in the dead leaves of the rainforest floor in eastern Brazil.

Nice to meet you!

THINK ABOUT IT

How does it make you feel to think that we might destroy the cure to a life-threatening disease without even knowing it?

DID YOU KNOW?

The biodiversity of rainforests may be 10 to 20 times as high as once thought.

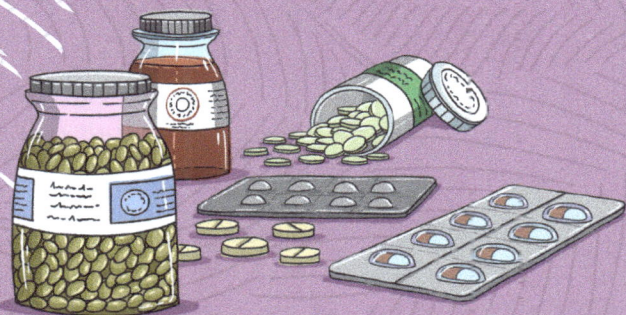

What would happen to rainforest plants and animals if the rainforests were destroyed?

- The rainforests are destroyed.
  - Trees and plants would be cut down.
  - Many animals and plants would be burned and killed by the fires set to clear the forests.
  - Some surviving animals would find new homes on the **grasslands** that replace the rainforests.
    - It would be hard for some rainforest species to survive on the grasslands without the right food or shelter. Their **population** would decrease.

Can you spot the leaf-tailed gecko? It uses camouflage to hide in trees from **predators**.

THINK ABOUT IT

Do you think that the leaf-tailed gecko's camouflage would work well if it had to live in grasslands instead? What would happen to it as a result?

GOING, GOING, GONE

Rainforest plants and animals would become **endangered** and eventually **extinct**.

Some animals would escape into nearby **habitats**.

Rainforest animals would have to compete for food with the animals that already live in these habitats. Both rainforest animals and the animals already in the habitat may struggle to find enough to eat, and their populations would decrease.

Mine

No, that's mine!

Spix's macaw is now considered extinct in the wild. However, zoologists have been breeding Spix's macaws in captivity in the hope that they can be reintroduced back into the wild.

Moving out

The rainforests are also home to **Indigenous** peoples, some of which have lived there for hundreds and even thousands of years. The destruction of the rainforests would make these people homeless and put them and their ways of life in serious danger.

There are lots of different groups of people around the world who call the rainforests their home, from the Yanomami of South America to the Dayaks of Southeast Asia and the Mbuti of central Africa. Many of these people make their living by hunting, fishing, collecting forest products, and farming. They know how to take from their rainforests without doing damage, and they often have their own cultures and traditions that have been passed down through the generations.

DID YOU KNOW? The Amazon rainforest is home to over 2 million Indigenous people spread across 400 different groups. Together, they speak over 300 indigenous languages.

A Yanomami man fishing

Can't catch me!

Over the past few hundred years, many Indigenous rainforest groups have lost their homes because of rainforest clearance. Even though nowadays many of them actually own the land they live on, developers threaten them and force them to leave.

If all the rainforests were destroyed, every Indigenous rainforest group would become homeless. Although people from the group would be able to find new homes outside of the rainforest, they wouldn't be able to live in the same way as before or practice their traditions. Over time, their culture would be lost forever.

These Mbuti boys in the Democratic Republic of the Congo are taking part in a traditional ceremony.

Moving out of the rainforests would put some Indigenous people's health at risk. Some groups haven't had the chance to be **vaccinated** against common diseases and have never been exposed to such illnesses as the common cold. If these people were forced to move into towns and cities, they would almost immediately catch diseases that could threaten their lives.

THINK ABOUT IT

How do you think Indigenous people would feel living in more built-up areas?

Other problems

Sounds bad, right? But wait, that's not all! The destruction of the rainforests will have a massive impact on our climate and the water cycle. Let's find out why.

As we've already seen, rainforests normally store huge amounts of carbon (see page 10). However, when rainforests are cut down, damaged, and burned, they turn into a carbon source and actually emit **carbon dioxide**. This is the total opposite of what our planet needs!

DID YOU KNOW?
Forest burning in the Amazon produces around three times more carbon dioxide than the forest can absorb.

HOW IS CARBON RELEASED FROM THE RAINFORESTS?

Loggers cut down trees.

Developers start fires to clear land.

Not this tree!

The forest catches fire.

Carbon is released from the rainforests.

Trees and plants die from lack of water.

The rainforest becomes drier, making it more likely to catch fire because of **wildfires** or fires set by humans.

The rainforest experiences **drought** because of **climate change**.

The more carbon that is released into our atmosphere, the worse **global warming** becomes. Higher temperatures on Earth lead to climate change, including more periods of drought and other extreme weather.

OTHER PROBLEMS

The burning of the rainforests also creates another nasty problem—air pollution. This can create breathing issues and long-term illnesses for people living nearby, especially those who are very young or old.

DID YOU KNOW? In 2019, more than 2,000 people in Brazil were hospitalized due to breathing problems caused by the burning of the Amazon rainforest.

The **grasslands** that would grow to replace the rainforests would absorb some carbon. However, they wouldn't be able to store nearly as much as the original rainforests. This would leave lots of extra carbon floating around our atmosphere, making the **greenhouse effect** worse.

The loss of the rainforests would also disrupt the water cycle. Without trees there to absorb and release water, there would be much less rainfall in the areas surrounding the rainforests. This could lead to **drought**. It's also likely that many places that were previously covered in rainforest would become much hotter without the cooling effect of rainforest trees releasing moisture from their leaves.

Many **crops** can't grow properly during droughts, which can lead to food shortages.

FUN FACT!
The moisture released from the leaves of just one tree each day has the same cooling effect on Earth as two household air conditioners.

I'm super cool!

Hi, we're Michael Wolosin and Nancy Harris. We bring together research from lots of different scientists to see how rainforests affect and are affected by **climate change**. It's clear that protecting our existing rainforests is one of the most important things we can do to reduce **global warming**. We hope that our research inspires governments and businesses to take the problem seriously and save our rainforests before it's too late.

Save our rainforests!

There is still hope for the rainforests, but we need to act now! The government and citizens of the countries where rainforests are found have the biggest role to play, but there are things that everyone around the world can do to help.

Above all other things, the most important thing we can do to save the rainforests is to stop all types of **deforestation.** To start with, all rainforests need legal protection that makes it illegal to cut down, burn, or destroy the plants and animals that live there. Many areas of rainforest are already legally protected, but we need to protect all of it to stand a chance of saving it.

DID YOU KNOW? Around half of the Amazon rainforest is protected land.

Protesters in Brazil demand that their government take better care of the Amazon rainforest.

If all the rainforests were under legal protection, it could be made illegal for anyone to destroy them. People who started fires or cut down trees would have to pay massive fines or even go to jail. This would discourage people from illegally destroying rainforest land.

Of course, we'd need to keep a very close eye on the rainforests to make sure that people were not breaking the rules. And that's easier said than done in such a huge area that's hard to patrol. Luckily, new technologies, such as drones and satellites, are making it easier than ever to monitor the rainforests. Not only is it much simpler and faster to gather data from the air, it's also much safer. Many people who illegally destroy the rainforests are criminals who can use violence to protect their businesses.

THINK ABOUT IT

What do you think the rainforest would look like from above? Draw a picture that shows wild forest and a deforested area.

These rainforest charity workers are using a drone to monitor an area of rainforest in Papua New Guinea.

Small changes to the items we eat and use daily can make a big difference to the rainforests! Two of the biggest offenders are palm oil and beef. Palm oil is in *so* many products and is very often grown on cleared rainforest land. Even if you don't live near South America, the beef that you buy in the supermarket may have been raised there on what was once a rainforest.

Buying these products puts money into the pockets of people who have already destroyed lots of the rainforest. It also shows them that there is a demand for their products, so they are more likely to continue destroying the rainforests to create more land for their farms.

If you can, look for products that don't contain palm oil, or switch to those that use palm oil that has been grown in a **sustainable** way. Choose local beef that has been raised on small farms, rather than on huge ranches in other countries.

Another way to be more rainforest-friendly is to ditch the beef altogether and eat more vegetables!

THINK ABOUT IT

How many products in your kitchen and bathroom contain palm oil? Can you find alternatives to any of them?

SAVE OUR RAINFORESTS!

Reducing the effects of **climate change** will also make a big difference to the rainforests. This can be achieved through big changes, such as switching from burning **fossil fuels** to using greener energy sources, such as solar and hydroelectric, as well as smaller individual choices, such as walking or taking public transportation rather than driving.

DID YOU KNOW?

If climate change continues, scientists believe that 30 to 60 percent of the Amazon rainforest will eventually turn into dry **grassland**, even without other forms of **deforestation**.

In order to protect rainforests in the long term, it's also important to support the communities that live there, so that they can make enough money to survive without having to clear rainforest land.

Sustainable farming provides income for local people and can also benefit the forest! Planting new plants and fruit trees can connect broken-up sections of the rainforest, creating a corridor for animals to travel along. Some **crops,** such as coffee and cacao, can even be grown inside the rainforest without clearing any land.

These people in Ghana are growing pineapples inside the rainforest.

Tourism is another way that rainforest communities can make a living. Local people can make money by working as tour guides, hotel staff, cooks, and cleaners. Being able to earn a living from the forest encourages people to protect it, since if the rainforest goes, so does their income!

SAVE OUR RAINFORESTS!

Hi! I'm Meg Lowman, an American scientist. Throughout my career, I've explored and studied many rainforests, with a particular focus on the canopy layer. It's hard work getting around in the canopy, as we can't swing from tree to tree like monkeys! I started building walkways to help me and my team explore. Now, we're building more canopy walkways in some of the most **biodiverse** rainforests on Earth, so that they can become destinations for research as well as ecotourism!

Although cleared rainforest land will eventually turn into **grassland** if left alone, it is technically possible to turn it back into a rainforest by replanting and reintroducing the species that once lived there. It's a very long and slow process, but worth it, because the new trees will help remove carbon from the atmosphere.

DID YOU KNOW?
Replanted rainforests are usually less biodiverse than original, wild rainforests.

Conclusion

We are closer than ever before to wiping rainforests off the face of our planet, thanks to **deforestation** and **climate change.** The loss of the rainforests would be a devastating blow to our planet. Millions of species would be gone forever, many people would be left without homes and a space to practice their traditions and cultures, and massive amounts of carbon would be released into the atmosphere.

DID YOU KNOW? In 2021, nearly 10 soccer fields worth of rainforests were lost every minute.

Please look after my home!

However, it's not too late to save the rainforests. If we act together as a planet to stop deforestation, change farming practices that damage the rainforest, and reduce the impact of climate change, we have the power to make a difference.

Planting wild trees and plants in cleared rainforest areas is a slow but effective way of regaining destroyed forests. But the best option is to save the rainforests we already have!

Above all, governments need to pass laws to protect the rainforests and prosecute those that break them. However, all of us can make small changes, for example, buying less beef, that will make a difference. If we work together, we can protect this unique and valuable **ecosystem** and preserve it for many more millions of years to come.

THINK ABOUT IT

What can you do to help save the rainforests? Talk to your friends and family about what they can do to help, too.

FUN FACT!

More than 100 world leaders have promised to end and start reversing deforestation by 2030. The clock is ticking!

Palm oil? No thanks!

Summary

So, what exactly would happen if all the rainforests were destroyed? Check your understanding of the information in this book.

- Many rainforest plants are dying because of **drought** linked to **climate change**.
- Huge areas of rainforest are being cleared for farms and development.
- Rainforest trees are cut down for their wood.
- Mines and **hydroelectric dams** are damaging and flooding areas of rainforest.
- **Grassland** would grow in the place of the cleared forests.
- **Crops** would be grown on some areas of grassland. However, the soil wouldn't be fertile enough to support the farms for long.
- Some grassland would be used as cattle ranches. The huge number of cows on the ranches would release large amounts of methane, a **greenhouse gas**.

All the rainforests are destroyed.

Many rainforest animals and plants would become **endangered** and eventually **extinct** without the right food and conditions.

Some rainforest animals would escape to other **habitats**. However, the arrival of new animals would upset the balance of these **ecosystems** and might put other animals there at risk.

Massive amounts of carbon would be released into the atmosphere.

Some rainforest animals would adapt to the new grasslands and would manage to survive.

More greenhouse gas in the atmosphere would make the **greenhouse effect** worse. This would result in more **global warming** and climate change.

Indigenous rainforest groups would lose their homes and would be forced to move to other areas.

Some Indigenous people could become very sick or even die if they were not given immediate treatment to protect them from diseases they've never encountered before.

The water cycle and rainfall patterns would be affected, which could lead to more periods of drought.

I'm so thirsty!

Many traditions and cultures would be lost.

Glossary

biodiverse—home to many different species of plants and animals

carbon dioxide—a gas absorbed by plants that can contribute to global warming

climate change—changes in the world's weather, in particular, an increase in temperature, which scientists believe are mainly due to human activity

crop—a plant grown for food, such as apples, carrots, or potatoes

decompose—to decay and break down into smaller parts

deforestation—the cutting down of trees and forests by people

drought—a long period with little or no rain

ecosystem—all of the living things in an area and the relationship between them

endangered—at risk of dying out because there are few of them left (oh no!)

equator—an imaginary line around the center of Earth

evaporation—when liquid turns into a gas, often because it is heated

extinct—an extinct animal or plant no longer exists on Earth because its entire species has died out, just like the dinosaurs

fossil fuel—a fuel such as natural gas, oil, or coal that was formed over millions of years from the remains of animals and plants

global warming—an increase in temperatures on Earth due to the greenhouse effect

grassland—a habitat mostly covered in grass

greenhouse effect—the effect caused by greenhouse gases (see below!)

greenhouse gas—a gas, such as carbon dioxide or methane, that gathers in the atmosphere and traps heat from the sun close to Earth's surface

habitat—the place where an animal or plant usually lives

herbivore—an animal that only eats plants for food, such as a rabbit

humid—when the air contains lots of water

hydroelectric dam—a dam that contains a machine that uses fast-flowing water to generate electricity

Indigenous—describes the people who originally lived in a place

nutrient—something that living things need in order to grow

photosynthesis—the process by which plants make their own food out of sunlight

population—how many animals or plants of the same type live in an area

predator—an animal that kills and eats other animals for food—watch out!

prey—an animal that is killed and eaten by other animals

purify—to remove bad substances from something

reproduce—to produce new, young animals or plants

sustainable—something that is able to continue over a long period of time because it doesn't damage the environment

transpiration—when water is lost from the surface of a plant (bye-bye!)

vaccinate—to put a substance into a living thing's body that protects them from disease by making them produce antibodies

wildfire—an out-of-control fire in a wild area

Review and reflect

COMPREHENSION QUESTIONS

Remarkable rainforests
- Rainforests are home to over half of the species found on Earth. Why are rainforests so biodiverse?
- How do rainforests help reduce global warming and climate change?

Rainforests in danger
- Rainforests are important ecosystems, so why are people still destroying them?

A new land
- Why are cattle considered one of the major contributors to global warming?
- Imagine the rainforests were destroyed and people instead used the land to grow crops. How might this practice actually further damage the land?

Going, going, gone
- How does looking at the cause and effect diagram on pages 26-27 help you understand the wide-reaching consequences of destroying all of Earth's rainforests?

Moving out
- In addition to plants and animals, who else calls rainforests home?
- How would destroying rainforests affect the Indigenous peoples living there?

Other problems
- Consider the cause and effect diagram on page 31. It doesn't seem to have a clear end. Rather, it shows a cycle. What does this mean for the condition of our rainforests?
- How would the loss of rainforests disrupt the water cycle?

Save our rainforests!
- What can governments do to help protect and save the rainforests?
- Who is Meg Lowman and how is she helping the rainforests?

Conclusion and summary
- After reading this book and considering what would happen if all the rainforests were destroyed, what is your biggest takeaway? Why?

MAKE A CHAIN OF EVENTS!

Creating a paper chain can help you explore and visualize how cause and effect relationships can be thought of as a sequence of events.

You'll need:
- Pencil
- Scratch paper
- Pens or markers
- Stapler and staples
- Strips of paper (2 colors, if possible)

Instructions:

1. **Select a focus:** Choose a specific aspect from the book that caught your attention—it could be something that has contributed to rainforest deforestation or something that would happen (or is already happening!) due to the destruction of rainforests.

2. **Brainstorm causes and effects:** On a sheet of scratch paper, brainstorm and list the causes and effects related to your chosen focus. Think critically about the factors that contributed to or resulted from your focus. You can always look back in the text for ideas!

3. **Write on strips:** Write each cause and each effect on its own strip of paper. If you have different colored paper, use one color for the cause strips and the other for the effect strips.

4. **Create the paper chain:** Organize your strips into causes and effects. Start forming a paper chain to show how a cause leads to an effect. Use the stapler to connect the two strips. Continue adding cause and effect strips as links in your chain. When you've finished, you should be able to start at the beginning of your chain and read through each chain link in a logical order.

5. **Linking multiple chains:** If your focus has multiple causes or effects, you can create additional chains and link them together to show how complex cause and effect relationships can be!

Write about it!

Look at the paper chain you created and how the causes link to effects (which in turn link to other causes!). How might breaking a link in the chain impact the overall sequence of events?

World Book, Inc.
180 North LaSalle Street
Suite 900
Chicago, Illinois 60601
USA

For information about other World Book publications, visit our website at www.worldbook.com or call 1-800-WORLDBK (967-5325).

For information about sales to schools and libraries, call 1-800-975-3250 (United States), or 1-800-837-5365 (Canada).

© 2024 (print and e-book) by World Book, Inc. All rights reserved. No part of this publication may be reproduced, stored in a retrieval system, or transmitted in any form or by any means (electronic, mechanical, photocopying, recording, or otherwise) without written permission from World Book, Inc.

WORLD BOOK and the GLOBE DEVICE are registered trademarks or trademarks of World Book, Inc.

Library of Congress Cataloging-in-Publication Data for this volume has been applied for.

What Would Happen If...?
978-0-7166-5448-3 (set, hc.)

All the Rainforests Were Destroyed?
ISBN: 978-0-7166-5453-7 (hc.)

Also available as:

ISBN: 978-0-7166-5459-9 (e-book)
ISBN: 978-0-7166-5465-0 (soft cover)

Staff

Editorial

Vice President
Tom Evans

Editorial Project Coordinator
Kaile Kilner

Curriculum Designer
Caroline Davidson

Proofreader
Nathalie Strassheim

Graphics and Design

Senior Visual Communications Designer
Melanie Bender

Digital Asset Specialist
Rosalia Bledsoe

Written by Izzi Howell
Illustrated by Paula Bossio

Developed with World Book by White-Thomson Publishing LTD
www.wtpub.co.uk

Acknowledgments

4-5	© Dmitry Rukhlenko, Shutterstock; © Eric Isselee, Shutterstock	22-23	© Petr Salinger, Shutterstock; © Avalon.red/Alamy Images
6-7	© Danita Delimont, Alamy Images; © saiko3p/Shutterstock	24-25	© neelsky/Shutterstock; © Joao Burini, Alamy Images
9	© Bill Roque, Shutterstock; © imageBROKER/Alamy Images; © EcoPrint/Shutterstock; © Papilio/Alamy Images	26-27	© Danny Ye, Shutterstock; © Jiri Balek, Shutterstock
10-11	© matthew25/Shutterstock; © Robert Harding, Alamy Images; © Irina Gutyryak, Shutterstock	28-29	© Nature Picture Library/Alamy Images; © Robert Harding, Alamy Images
12-13	© Thammanoon Khamchalee, Shutterstock; © jack-sooksan/Shutterstock; © PhotoStock-Israel/Alamy Images; © kakteen/Shutterstock; © imageBROKER/Alamy Images	30	© RDW Aerial Imaging/Alamy Images
		32-33	© Roger Tillberg, Alamy Images; © juerginho/Shutterstock
14-15	© inga spence/Alamy Images; © Ody Stocker, Shutterstock; © Rich Carey, Shutterstock	34-35	© Hemis/Alamy Images; © Cintia Erdens Paiva, Alamy Images
16-17	© Tristan Tan, Shutterstock; © PARALAXIS/Shutterstock	36-37	© gowithstock/Shutterstock; © Aleksandar Tomic, Alamy Images
18-19	© Africa Studio/Shutterstock; © Yogie Hizkia, Shutterstock	38-39	© Ammit Jack, Shutterstock; © Ron Giling, Alamy Images
20-21	© TFoxFoto/Shutterstock; © Peterson IMG/Shutterstock; © alvarobueno/Shutterstock; © GoodFocused/Shutterstock	40-41	© imageBROKER/Alamy Images; © AustralianCamera/Shutterstock
		46-47	© Roi and Roi/Shutterstock